SPACE ...ES

...G THE ...STEM

PETER GREGO

QED Publishing

QED

First published in the UK in 2007 by
QED Publishing
A Quarto Group company
226 City Road
London EC1V 2TT
www.qed-publishing.co.uk

Reprinted in 2008

A catalogue record for this book is available from the British Library.

ISBN 978 1 84835 013 7

Written by Peter Grego
Produced by Calcium
Editor Sarah Eason
Illustrations by Geoff Ward
Picture Researcher Maria Joannou

Publisher Steve Evans
Creative Director Zeta Davies
Senior Editor Hannah Ray

Printed and bound in China

Picture credits
Key: T = top, B = bottom, C = centre, L = left, R = right, FC = front cover, BC = back cover

Corbis/Eugen/Zefa 9, /NASA 8–9, /Robert Weight/Ecoscene 11T; **CTIO**/NOAO/AURA/NSF 7B; **Getty Images**/Amana 14B, /Iconica 11B, /Photodisc FCT, FCC, 8, 12, 18, 21T, 27B; **NASA** 1–5, 10, 13B, 17T, 17B, 19T,19B, 21B, 22, 25T, 25B, 28, BC, /ESA/M. Robberto and the Hubble Space Telescope Orion Treasury Project Team 6, /The Hubble Heritage Team (STScI/AURA) 24, /JPL 13T, 14T, 15, /JPL-Caltech 7T ; **Peter Grego** 23B; **Science Photo Library** 24, /Harvard College Observatory 29, /Mark Garlick 20, FCB, /NASA 22–23, /NASA/ESA/STSCI/Erich Karkoschka/University of Arizona 26, /Sheila Terry 27T, /Detlev Van Ravenswaay 28–29; **Wikipedia Commons Public Domain** 5.

Words in **bold** can be found in the Glossary on pages 30–31.

Contents

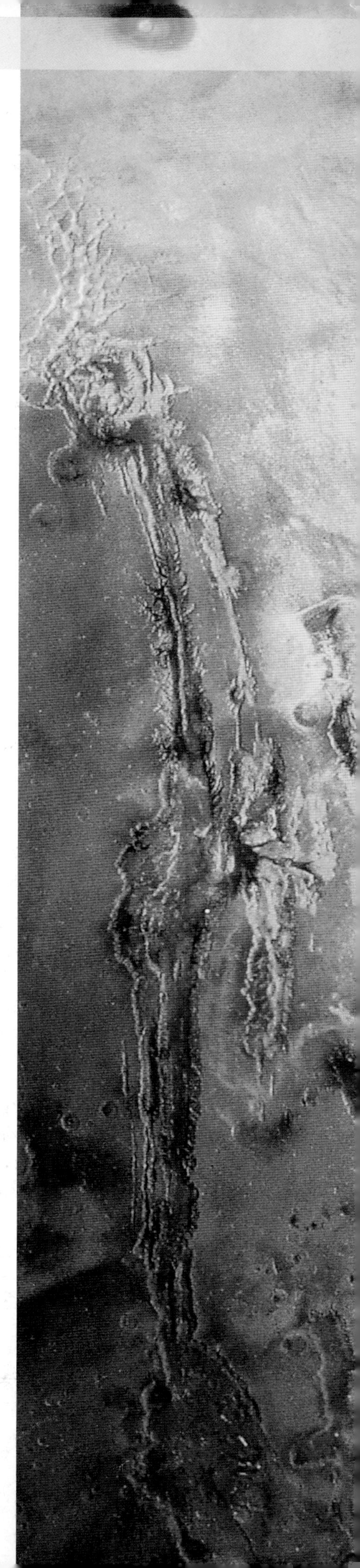

Our cosmic backyard

Our **planet**, Earth, is one of eight planets that make up an area of **space** we call our **Solar System**. Inside the Solar System are our **Sun** and **Moon**. It is also home to the eight planets, Mars, Venus, Mercury, Earth, Jupiter, Saturn, Uranus and Neptune, their moons, and many **comets** and **asteroids**.

The four inner planets

The four planets nearest the Sun are Mercury, Venus, Earth and Mars. These are called the four 'inner planets'. Each has a solid surface and, except Mercury, is surrounded by an **atmosphere**. Only the Earth's atmosphere is suitable for us to breathe.

The four outer planets

The four 'outer planets' are Jupiter, Saturn, Uranus and Neptune. These are giant balls of mainly **hydrogen** and **helium gas** and are the planets furthest from the Sun.

The four inner planets of our Solar System: Mercury, Venus, Earth and Mars.

The dwarf planets

As well as the eight major planets, a few smaller solid balls orbit the Sun. These are called '**dwarf planets**'. They include the largest asteroid, Ceres, and the frozen worlds of Pluto and Eris.

Venus

Biography

Johannes Kepler (1571–1630) and his planetary laws

Four hundred years ago, the German astronomer Johannes Kepler worked out that all planets in the Solar System travel around, or orbit, the Sun in an oval-shaped path called an 'ellipse'. The nearer a planet moves towards the Sun, the faster it travels. As the planet moves away from the Sun, it slows down. This rule is the same for all orbiting objects – a moon orbiting a planet or a comet orbiting the Sun.

Johannes Kepler.

Comets

Comets are mountain-sized, dirty snowballs. They are mainly found in the far-out depths of space, great distances from the Sun. Whenever their orbit takes them close to the Sun, they heat up and throw off clouds of dust and gas.

Earth

Formation of the Solar System

Around 4.6 billion years ago, scientists believe that something amazing happened in our **galaxy**, the Milky Way. A star exploded, sending a ripple through a large cloud of nearby dust and gas. The ripple squashed the cloud together, making it thicker in parts.

Gravity pulled on the thicker areas, which collapsed in on themselves. As the cloud collapsed, it began to spin faster and faster. This made it become flatter, thinner and cooler towards the edges, and hotter at its centre. Clumps of material in the cloud then began to stick together. As these areas became bigger, gravity caused them to attract other clumps of material. The clumps of material became bigger and bigger, and eventually grew into the planets, moons, asteroids and comets that make up our Solar System.

The Sun was born inside a big cloud of gas and dust, called a nebula.

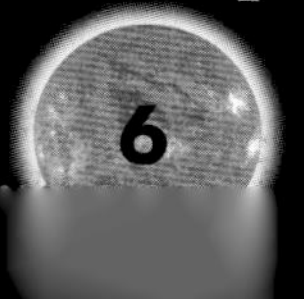

The young Sun was surrounded by clouds of dust and gas, which clumped together to create the planets.

Birth of the Sun

The temperature at the centre of the of the collapsing Milky Way gas cloud became so high that it caused a huge reaction, called a **nuclear reaction**, which created the Sun.

The incredible heat and energy made by the newborn Sun cleared away the gases and ice in the inner part of the Solar System. In this gas- and ice-free area are the four solid inner planets. Gas and ice remained in the outer system, which is where the four outer planets (made of gas) are found. The very edge of the Solar System is still freezing and here millions of icy comets exist.

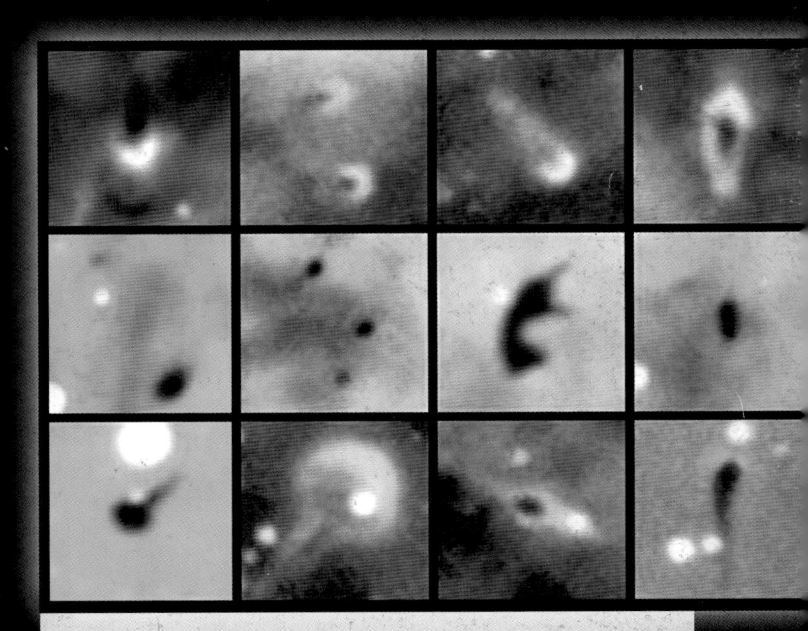

This collection of odd looking objects was found in the Carina Nebula (within the Milky Way) by a very powerful **telescope** called the Hubble Space Telescope. Each one is thought to be a very young solar system in the making.

Our Sun, the nearest star

The Sun is a star that shines in the same way as any other star – by changing its hydrogen gas to helium gas, which makes light and heat. It is an enormous ball made up of mainly hydrogen gas, which is found throughout the **Universe**. The Sun is incredibly hot – in fact, the **pressure** and heat at the centre of the Sun is a hundred thousand times hotter than your oven!

Hurricane

All weather on Earth is driven by the Sun's heat.

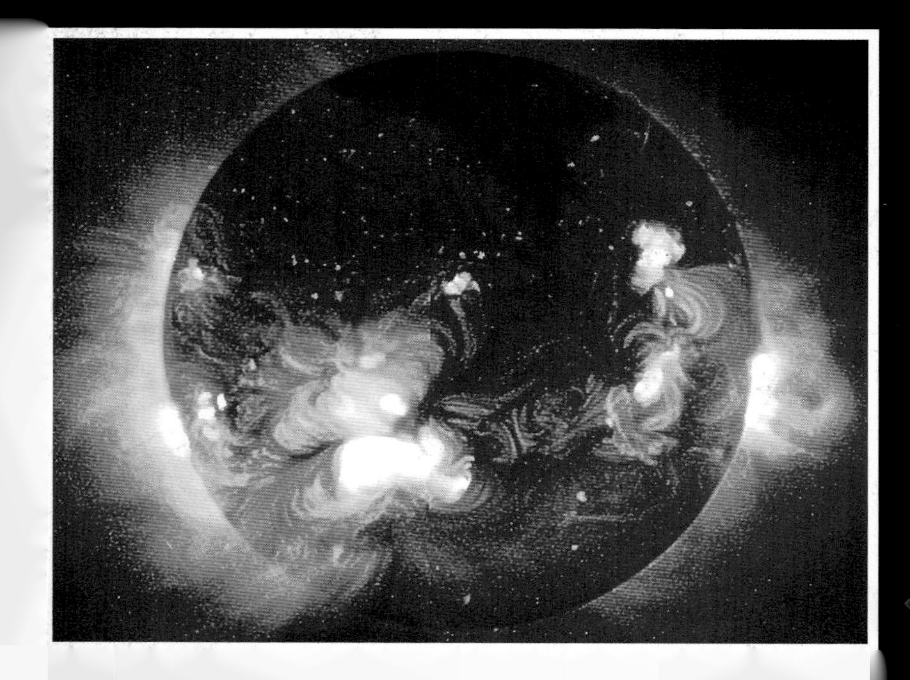

The Sun is a giant, glowing ball of gas

All plant and animal life on Earth depends upon energy from the Sun.

Slowly dying

The Sun provides Earth with heat and light, which drives our planet's weather and allows life to exist. If the Sun were somehow switched off, our planet would become a frozen ball upon which only the simplest life forms could survive.

Over the last 4.6 billion years, the Sun has been slowly using up its hydrogen gas, which fuels its heat. The Sun is now about halfway through its life. Eventually, in the distant future, it will use up all of its fuel and die.

!

Amazing

Burning up energy

Every second, the Sun loses four million tonnes of its weight as it burns up the gas at its centre. Despite this, the Sun will carry on shining for several billion more years.

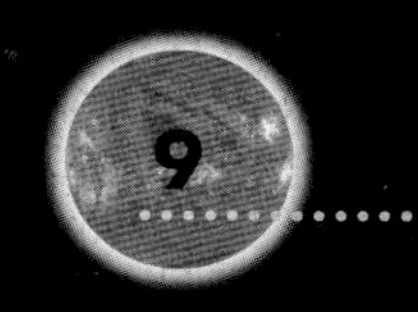

Solar spectacles

Astronomers use special telescopes to look at a very hot area above the Sun's surface. Here they can see gigantic flames of hot gases, called **solar** flares and prominences. Sunspots can also be seen on the Sun's surface. These are dark spots that are actually very hot and bright, but seem dark because the rest of the Sun's surface is even hotter and brighter. Sunspots are caused when hot gases flow around **magnetic fields** at the Sun's surface, leaving a slightly cooler, darker-looking central area.

Warning!

Never look directly at the Sun with your eyes, and never use binoculars or a telescope to view it. Raw sunlight can damage your eyes and could even make you blind. Astronomers study the Sun safely using special telescopes and equipment.

A total **eclipse** of the Sun.

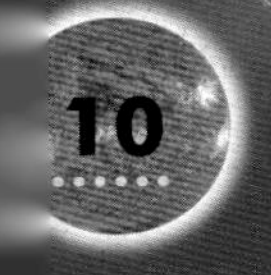

These bright blobs in the sky are called 'sundogs'. They are caused when sunlight is bent by ice crystals high in the atmosphere, producing a bright spot in the sky on either side of the Sun.

Total sunblock!

Sometimes the Moon moves between the Earth and the Sun. This blocks the Sun's light and causes an eclipse. Most eclipses are just partial (do not cover the Sun completely). Occasionally the Moon completely covers the Sun for a few minutes, causing a total eclipse. During a total eclipse, the sky goes dark, bright stars and planets are clear and the edge of the Sun can still be seen around the dark Moon.

Light fantastic

Aurorae are fantastic, multicoloured lightshows that appear in Earth's atmosphere. They are caused by winds of energy that blow from the Sun. Aurorae are best seen from **polar areas** because the winds from the Sun travel along the magnetic fields above the Earth's **poles** and cause gases high in the atmosphere to glow. Large aurorae can sometimes be seen from as far south as the Mediterranean!

This bright green aurora was seen in the skies above western Iceland.

Mercury

Mercury is the planet closest to the Sun. It orbits the Sun incredibly quickly, making four complete journeys around it every year. Because Mercury is so close to the Sun, it never moves far from its glare. It moves so quickly that it can only be seen from Earth six times a year, for two weeks at a time before sunrise or after sunset.

Smallest planet

Mercury is the smallest of all the planets. It has a very thin atmosphere and has no weather at all. With no atmosphere to spread the heat around the planet, there is a huge difference in temperature between its day and night sides. Its Sun-facing day side becomes as hot as an oven at its hottest setting, while its night side plunges to around twice as cold as the coldest temperature ever recorded on Earth.

Mercury, photographed by the Mariner 10 space **probe**.

A close-up of Mercury's surface, showing the Caloris **crater** near the shadow at the left.

Enormous crater

Packed with craters, Mercury's rocky surface looks a lot like the Moon. The craters were caused by asteroid impacts – city-sized chunks of rock that smashed into the planet's surface. Most of Mercury's craters were made many billions of years ago, shortly after the Solar System formed. The planet's biggest crater is a giant scar that was caused by an asteroid. It is called Caloris and the whole of England could fit inside it!

When it moves directly between the Sun and the Earth, Mercury is seen as a tiny black dot. This picture shows its path across the Sun during the course of the evening of 8 November 2006.

19:15
20:15
21:15
22:15
23:15
00:00

Amazing

Rare sightings

On rare occasions, Mercury moves directly between the Sun and the Earth, when it can be seen through a telescope for just four hours as a black silhouette**. The next time Mercury can be seen in this way will be on 9 May 2016.**

Venus

Venus is the second planet from the Sun and is nearly as big as the Earth. The Romans named Venus after their goddess of love because it is such a bright and beautiful planet. In fact, it is so bright that it is easy to see in the sky, and has even been mistaken for a UFO!

Seeing with radar

Venus is covered with a very cloudy atmosphere. This reflects a lot of sunlight back into space and is what makes the planet so bright. Although the planet's clouds always hide its surface from view, scientists have been able to see below them with space probe **radars**.

A view of Venus through a telescope.

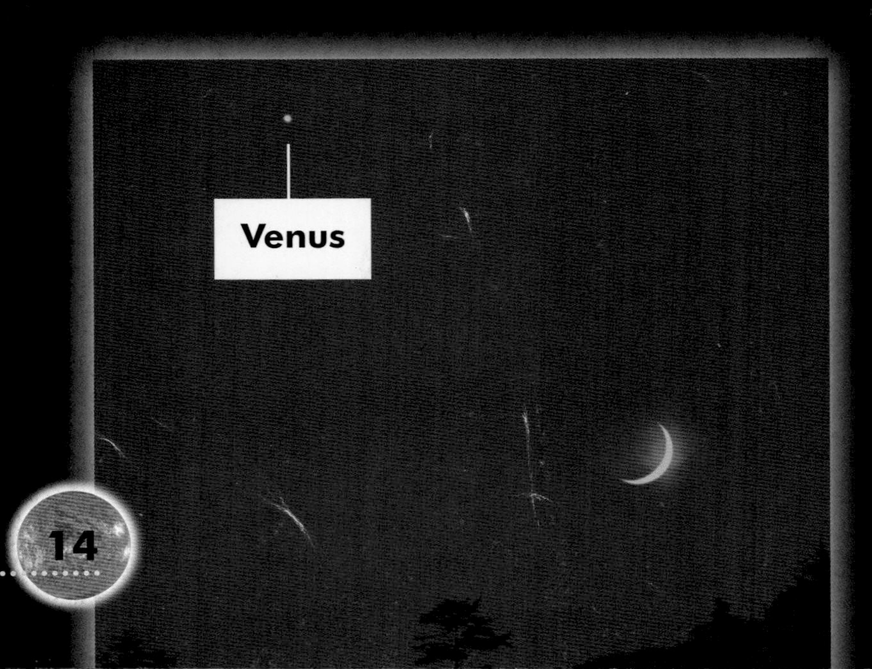

Venus and the Moon can be seen shining brightly in the evening sky.

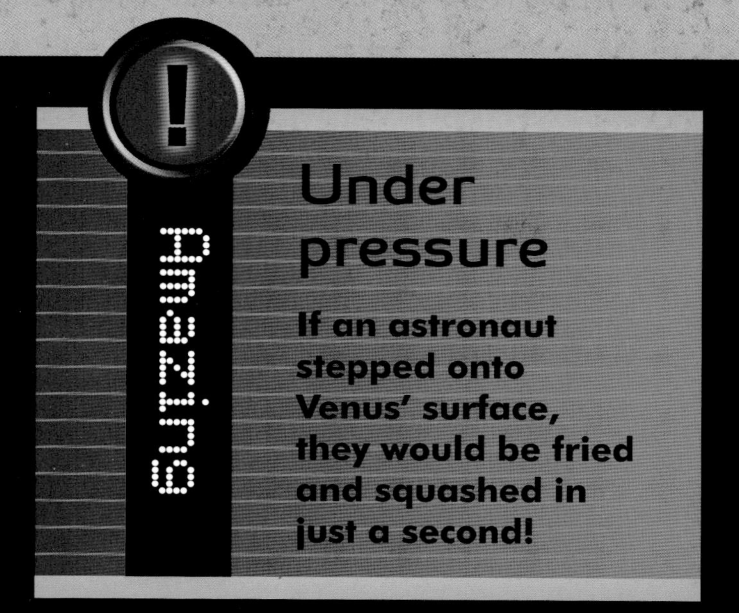

Probing for information

Several probes have landed on Venus and taken pictures. The planet's surface is blisteringly hot and the pressure of its atmosphere is enormous – equal to 1000m below sea level! As a result, the probes did not survive for long.

Mountainous landscape

Most of Venus is covered by huge, rolling plains. From these rise a few large **continent**-sized regions, topped with enormous mountain ranges. Venus's highest mountain, Maxwell Montes, is more than 3km taller than Mount Everest, the Earth's highest peak. The planet's largest continent is called Aphrodite Terra and is about the same size as Africa. The planet may have active volcanoes today, and any craters caused by asteroids were certainly worn away by its atmosphere and volcanic eruptions in the past.

Beneath its clouds, Venus has a wonderful landscape of rolling plains, volcanoes and mountains.

Earth and Moon

Planet Earth is the largest of the four inner planets. Along with its **satellite**, the Moon, it orbits the Sun once a year. The surface of the Earth is called the crust, and is a thin layer of rock. It includes the continents and the ocean floor. Beneath the Earth's surface it is so hot that solid rock melts into a flowing liquid called **magma**. Magma continually churns beneath the Earth's surface and pushes on the solid crust above it.

Moving plates

The Earth's crust is a patchwork of 12 large **plates**, which slowly move against each other. When the plates collide with each other, they crumple. This pushes the Earth's surface upwards to create a mountain range, such as the Himalayas. Where thin ocean crust is forced beneath thicker continental crust, the rocks melt deep inside the hot magma. This melted rock rises like a bubble into the crust above. Volcanoes occur where it breaks through the Earth's surface.

The Earth seems to be a giant jigsaw when its plates are drawn on a map.

The beautiful, blue planet Earth, seen from space.

Water world

Key Concept

The Earth is virtually covered with water – only a quarter of the surface is dry land. To date, it is the only place in the Universe where we can be certain life exists.

The Moon

The Earth's Moon is a lifeless place, covered in asteroid craters.

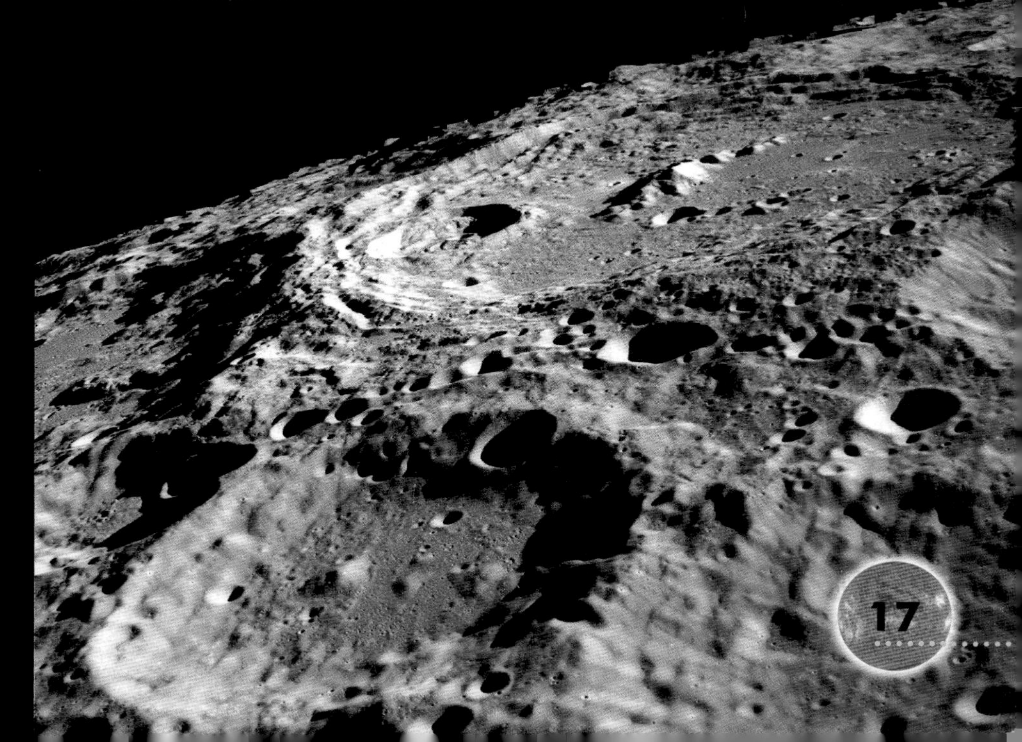

The Moon is almost as wide as the USA. It orbits the Earth once a month. No wind ever blows there, no clouds ever appear in its skies and no water ever quenches its dry surface. Life never developed on the Moon. Some parts of the Moon's surface are covered with asteroid craters. There are also many large plains of dark solid, **lava** known as 'seas'.

Mars

Mars is just over half the size of the Earth and is the fourth planet from the Sun. It takes nearly two years to orbit the Sun. Every few years the planet appears particularly bright for a month or two in the midnight sky, when it is at its closest point to the Earth. Believing the planet looked like a drop of blood, the Romans named it Mars after their god of war. Its red colour is because of the rust in its **iron**-rich soil.

Mars, viewed by the Mars Global Surveyor.

Mariner Valley

Amazing

Martians!

Astronomers in the 19th and early 20th centuries imagined they had seen a network of straight lines on Mars. Some claimed they might be canals **built by intelligent Martians!**

Cold, dusty planet

Mars is very cold. It has just a thin atmosphere in which we would not be able to breathe. However, of all the planets in the Solar System, Mars is the most similar to Earth. A day on Mars is just a little longer than our own day. The planet also has its own seasons.

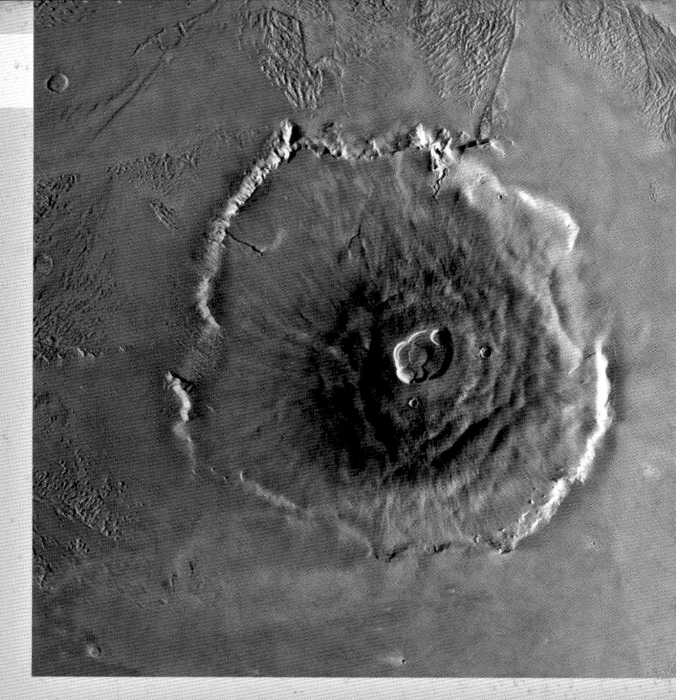

Mount Olympus is a giant extinct volcano that towers above the plains of Mars.

Rocky deserts and dust **dunes** spread across Mars, and its poles are covered with ice caps. Mars has **canyons**, such as the Mariner Valley, and **gullies**, which show that large amounts of water may once have flowed across its surface when the planet was a lot warmer and wetter than it is today.

Life on Mars?

It is possible that **primitive** life may have existed on Mars a long time ago. In August 1996 **NASA** claimed it had discovered traces of **fossil bacteria** in a **meteorite** from Mars. There may even be life on Mars today, hidden in its red soil.

NASA has landed several probes on the surface of Mars, including two robotic, wheeled rovers called Spirit (left) and Opportunity. These two probes discovered that certain rocks and minerals on the planet had formed in water, proving that there was once water on Mars.

Asteroids and dwarf planets

In the 18th century, astronomers discovered a gap in space between Mars and Jupiter. Believing a mysterious planet existed there, they formed a group to search for it. The group was called the 'Celestial Police'. On 1 January 1800, astronomer Giuseppe Piazzi spotted a faint object slowly orbiting the Sun between Mars and Jupiter. It was so small that it simply looked like a point of light, so it was said to be a '**minor planet**' or asteroid. It was later named Ceres.

As wide as Texas, Ceres is the biggest object in the Asteroid Belt.

Minor planets

Ceres turned out to be just one of thousands of minor planets orbiting the Sun in what we now call the **Asteroid Belt**, between Mars and Jupiter. Over 30 000 minor planets have been found to date.

Asteroid zone

Asteroids are chunks of rock that formed during the early days of the Solar System. Groups of asteroids orbit in front of and behind Jupiter. An asteroid zone also lies further out between Jupiter and Neptune.

Large chunks of rock called meteors sometimes blaze through our skies and hit the Earth's surface, causing craters.

Amazing

Smashing rocks

Small asteroid chips sometimes land on Earth. They are called meteorites. Sometimes very large chunks of asteroid head our way and blast out craters in the Earth's crust. 65 million years ago an asteroid impact is thought to have wiped out the dinosaurs.

Asteroid

The first dwarf planets

At 1000km wide, Ceres is the largest object in the Asteroid Belt. In 2006, Ceres was upgraded from an ordinary asteroid to a 'dwarf planet'. Unfortunately, the same meeting decided to downgrade Pluto from a regular planet (something it had been considered to be since its discovery in 1930) to a dwarf planet. Measuring just 2390km across, Pluto was thought to be too small to be a major planet. Another dwarf planet named Eris, slightly larger than Pluto, was discovered in 2003.

The asteroid Eros is a strangely shaped chunk of rock that is about as big as the island of Manhattan, in New York, USA.

Jupiter

Jupiter is the Solar System's biggest planet – it is so large that more than a thousand Earths could fit inside it! It is made up mainly of hydrogen and helium gas.

Ball of gas

Jupiter spins on its **axis** once every ten hours – so fast that it bulges out at its centre. Its cloudy atmosphere has light- and dark-coloured streaks, amazing cloud patterns and spots. In its atmosphere, Jupiter also has a storm that is larger than Earth! It is called the Great Red Spot and may have swirled around Jupiter's atmosphere for over 350 years. Because Jupiter is a ball of gas, it does not have a solid surface where a space probe could land. At the planet's centre there may be a core of **molten rock** bigger than Earth.

Jupiter's moons

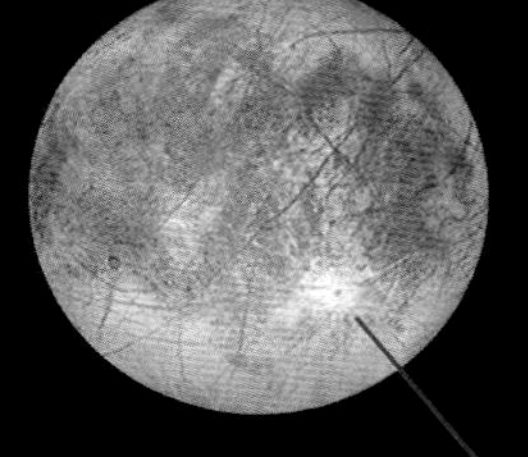

Jupiter has more than 60 moons. The two largest moons, Ganymede and Callisto, are bigger than the planet Mercury. Io is bigger than our own Moon and is covered in active volcanoes. Europa is a little smaller. Scientists think that beneath its icy surface there may be an ocean of warm, salty water in which primitive sea life has developed.

Io

Ganymede

Callisto

Great Red Spot

Jupiter's Great Red Spot can be seen on its surface.

Project

Viewing Jupiter and its moons

Use a good astronomy **magazine to find the position of Jupiter in the night sky. Then look at the planet through binoculars. You will probably see a small bright oval with several dimmer points of light near to it. If you draw the position of these points of light, and do the same again the following night you will find that they have moved. These are Jupiter's moons.**

Jupiter and its four big moons, Ganymede, Callisto, Io and Europa.

Saturn

Saturn is the Solar System's second-biggest planet. A ball of mainly hydrogen and helium gas, like Jupiter it bulges outwards at its centre because it spins so quickly on its axis. It is the least **dense** of all the planets – if it was shrunk to the size of your fist it would weigh less than a snowball!

Windy planet

Saturn's atmosphere has a few streaky clouds, but it is much calmer than stormy Jupiter. The planet has very high winds, which can be five times faster than the strongest hurricanes on Earth!

Saturn is famous for its beautiful **rings** of rock and ice.

Saturn's system of rings is one of the Solar System's most beautiful sights. This close-up view by the Cassini probe shows that the rings are made up of hundreds of narrow ringlets.

Rings of rock and ice

From a distance, Saturn's rings look like solid flattened hoops. They block out the light from the Sun to cast dark shadows onto the planet below. The rings are made up of millions of small chunks of rock and ice. Viewed from the Earth over the years, the rings appear to open up, close and open up once more. This is caused by the changing tilt of Saturn as it orbits the Sun.

Saturn's moons

Around 60 large moons orbit Saturn. One of them, Titan, is a true giant. Bigger than the planet Mercury, Titan is the only satellite in the Solar System that has its own atmosphere. Titan has an icy landscape of hills and volcanoes. It may also have rivers and lakes made up of a **chemical** called methane.

Titan, Saturn's biggest Moon. Titan has a yellow surface made of nitrogen and **methane gas**. The yellow surface is hidden behind a veil of blue-green cloud.

Uranus and Neptune

Of all the eight planets in our Solar System, Uranus and Neptune exist furthest from the Sun. Uranus takes more than a human lifetime to orbit the Sun. It is green-blue in colour and four times wider than Earth. It was the first planet ever discovered through a telescope. Until then only Mercury, Venus, Mars, Jupiter and Saturn were known (the only planets visible to the naked eye). Uranus has many icy satellites, the biggest of which is Titania.

This photograph of Uranus taken by the Hubble Space Telescope shows the planet's clouds, delicate rings and some of its small moons.

Moons

Mysterious planet

During the early 19th century, astronomers discovered that Uranus took an unusual path around the Sun. This could only be caused by another planet, further out in space, tugging on it to change its orbit. In 1846 this mysterious planet was seen for the first time and named Neptune, after the Roman god of the sea.

Biography

William Herschel (1738–1822) discovers Uranus

On 13 March 1781, amateur astronomer William Herschel spotted a small circular object that didn't look like a star or comet. It was a mysterious new planet that lay far beyond Saturn. Saturn is the Roman name for the Greek god Cronus. The new planet was named Uranus, in honour of Cronus's father.

William Herschel. The chart he is holding shows his discovery of Uranus.

Neptune

Neptune is the most distant planet in our Solar System. Its orbit takes it 30 times further from the Sun than Earth's. On 29 May 2011 Neptune will have made just one circuit around the Sun since its discovery in 1846.

Even though Neptune is far out in freezing space, a lot of activity takes place there. Storms often well up in the planet's cloud belts, blown by the strongest winds in the Solar System.

Neptune's biggest moon, Triton, is a frozen icy world, with active icy volcanoes that spurt out nitrogen gas.

This picture of Neptune was taken by the Voyager 2 space probe in 1987.

Pluto, Eris and the Kuiper Belt

When tiny, far-off Pluto was discovered in 1930, it was declared the Solar System's ninth planet. Pluto's orbit takes it 50 times further than the earth from the Sun and takes almost 250 years to complete! Pluto is a little smaller than our own Moon and is one of the few large objects in the Solar System still to be seen by a space probe. Pluto has a big satellite called Charon, which is more than half its size. It also has two tiny moons called Nix and Hydra. Even further from the Sun than Pluto is the Oort Cloud, an area of space filled with comets.

The Kuiper Belt

Hundreds of big asteroids exist beyond Neptune in a zone called the Kuiper Belt. In 2003 a Kuiper Belt object slightly bigger than Pluto was discovered and named Eris. Astronomers recently decided that both Pluto and Eris should be classed as 'dwarf planets', but were not big enough to be major planets. Our Solar System now has just eight planets.

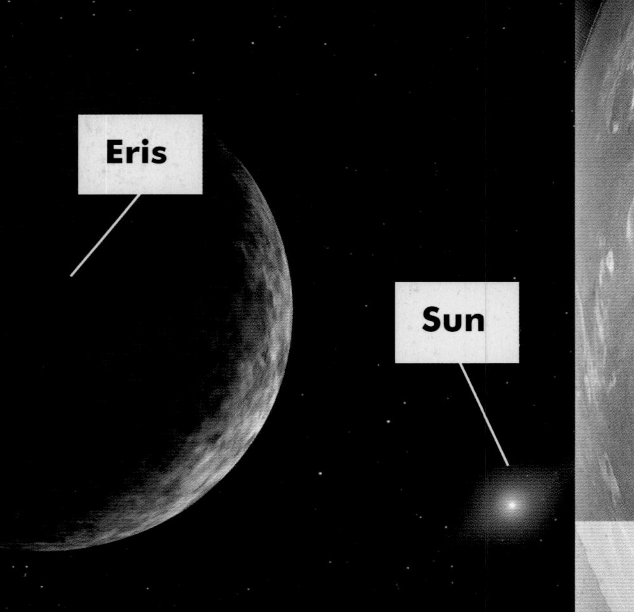

Eris is so far away that the Sun looks like a bright star in its skies.

A space probe took this photograph of Halley's Comet. It has an icy surface, covered in craters, and is surrounded by gas and dust.

Comets

Comets are mountains of ice, dust and rock, which formed in the deep, frozen areas of the outer Solar System. A number of comets travel regularly to the inner Solar System. The most famous of these is Halley's Comet, which approaches the Sun every 76 years. Many billions of comets may exist in the Oort Cloud.

Pluto's moon, Charon.

Dwarf planet Pluto.

Key Concept

Comet tails

When a comet approaches the Sun, its icy surface heats up and gives off gases. As the ice melts, bits of dust and rock that were trapped in it are freed. They float into space and trail behind the comet, creating an amazing tail.

Glossary

asteroid a lump of rock in space
Asteroid Belt a band of space between Mars and Jupiter in which thousands of large asteroids are found
astronomy the scientific study of objects in space
atmosphere a mixture of gases found around a star or a planet
aurorae coloured glows in the night sky made when energy from the Sun causes gases in the atmosphere to light up
axis the line around which a planet turns
canal an artificial waterway
canyon a steep-sided valley
chemical a substance that is created when atoms change
comet a city-sized 'dirty snowball' in space. When a comet gets near the Sun, it heats up and produces a long tail of gas and dust
continent a large land mass. Earth has five continents
crater a circular, bowl-shaped hole in a planet's surface. Craters are blasted out by the impact of an asteroid, or the explosion of a volcano
dense something having a high weight compared to its volume
dune a wind-blown pile of sand
dwarf planet an object orbiting the Sun with enough gravity for it to have pulled itself into a rounded shape, but too small to be considered a proper planet
eclipse when the Moon moves between the Sun and the Earth and it covers the Sun's light so that only the edge of the Sun can be seen
fossil bacteria the impression in rock of something that lived long ago
galaxy a collection of millions, hundreds of millions or billions of stars, all held together by gravity
gas a chemical which is not liquid or solid
gravity a force that acts throughout the Universe. The Earth's gravity holds you to its surface, and the Sun's gravity holds the Earth in its orbit. The bigger and more massive the object, the more gravity it has
gullies small water-cut valleys
helium a very lightweight gas
hydrogen a very light and colourless gas
iron a metal

lava hot, melted rock which has bubbled up from below a planet's surface (usually through a volcano)

magma molten rock beneath a planet's crust

magnetic fields areas of magnetic energy

meteorite a small rock that has fallen from space onto Earth

methane gas a smelly gas

minor planet an asteroid

molten rock rock so hot that it is melted and is runny

Moon the Moon is the Earth's only natural satellite

NASA National Aeronautics and Space Administration, the United States' national space agency

nuclear reaction a burst of energy that is made when atoms hit each other at high speeds

planet a large, round object that orbits a star. The Sun has eight major planets – Mercury, Venus, Earth, Mars, Jupiter, Saturn, Uranus and Neptune

plate a large portion of the Earth's crust floating on the magma

polar area the regions around a planet's poles

pole the point on the surface of a planet around which everything revolves

pressure the amount of force applied by a substance on another

primitive very simple and undeveloped

probe an automated spacecraft which gathers scientific information

radar a sensing system that uses radio waves to detect physical objects

rings large, flat circular hoops made up of dust, rubble and/or ice chunks. Rings surround all four gas giants, but are particularly bright and beautiful around Saturn

satellite an object that orbits around a larger object. Satellites can be natural, such as moons, or man-made, such as machines that orbit Earth and send back information

silhouette a dark, shadowed shape seen against a brighter background

solar anything to do with the Sun

Solar System an area of space containing our Sun, the planets and their moons, asteroids and comets

space everything outside the Earth

Sun our nearest star. The Sun is a huge ball of gas

telescope an instrument used by astronomers to study objects in space

Universe everything that exists

Index